Layers of the Rain Forest

by Julie Murray

Dash!
LEVELED READERS
An Imprint of Abdo Zoom • abdobooks.com

Level 1 – Beginning
Short and simple sentences with familiar words or patterns for children who are beginning to understand how letters and sounds go together.

Level 2 – Emerging
Longer words and sentences with more complex language patterns for readers who are practicing common words and letter sounds.

Level 3 – Transitional
More developed language and vocabulary for readers who are becoming more independent.

abdobooks.com

Published by Abdo Zoom, a division of ABDO, PO Box 398166, Minneapolis, Minnesota 55439. Copyright © 2023 by Abdo Consulting Group, Inc. International copyrights reserved in all countries. No part of this book may be reproduced in any form without written permission from the publisher. Dash!™ is a trademark and logo of Abdo Zoom.

Printed in the United States of America, North Mankato, Minnesota.
102022
012023

Photo Credits: Getty Images, Minden Pictures, Shutterstock
Production Contributors: Kenny Abdo, Jennie Forsberg, Grace Hansen, John Hansen
Design Contributors: Candice Keimig, Neil Klinepier

Library of Congress Control Number: 2022937235

Publisher's Cataloging in Publication Data

Names: Murray, Julie, author.
Title: Layers of the rain forest / by Julie Murray
Description: Minneapolis, Minnesota : Abdo Zoom, 2023 | Series: Rain forest life | Includes online resources and index.
Identifiers: ISBN 9781098280109 (lib. bdg.) | ISBN 9781098280635 (ebook) | ISBN 9781098280932 (Read-to-Me ebook)
Subjects: LCSH: Forests and forestry--Juvenile literature. | Rain forests--Juvenile literature. | Temperate rain forest ecology--Juvenile literature. | Biotic communities--Juvenile literature.
Classification: DDC 577.34--dc23

Table of **Contents**

Layers of the Rain Forest

Rain forests have four layers. Each layer is like a new world. The plants, animals, and weather are different in each layer.
Emergent Layer
Canopy Layer
Understory Layer
Forest Floor

The Four Layers

The topmost, or emergent, layer is made up of all trees. This layer gets the most sunlight and weather. Branches are thin and the leaves are **waxy**.

Many animals in the
emergent layer are fliers,
like birds and butterflies.
Pygmy gliders also live here.

The canopy is the next layer. It is home to most of the rain forest's life. There is thick plant growth. Many trees here produce fruits.

Many animals, like monkeys and sloths, live in the canopy layer. It is warm, safe, and has lots of food.

The third layer down is the understory. It is home to small trees, young trees, and **shrubs**. Many plants here have large leaves to capture the little sunlight that shines.

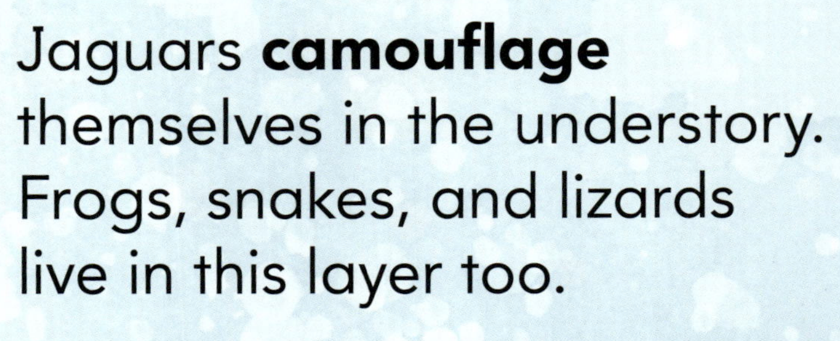

Jaguars **camouflage** themselves in the understory. Frogs, snakes, and lizards live in this layer too.

The last layer is the dark forest floor. Not many plants grow here. **Leaf litter** covers the ground.

Slugs, worms, and beetles live on the forest floor. They feed on dead plants and animal droppings. Bigger animals like elephants and gorillas live here too!

More Facts

- The treetops in the emergent layer are 200 feet (61 m) tall.

- The canopy blocks most of the sun, rain, and wind from reaching the lower layers.

- The forest floor receives less than 2% of the canopy light.

Glossary

camouflage – to hide by coloring or covering to look like the surroundings.

leaf litter – decomposing but recognizable leaves and other debris forming a layer on top of the soil, especially in forests.

shrub – a plant with woody stems that branch out close to the ground.

waxy – marked by or covered in a smooth, shiny, and wax-like texture. In leaves, having a wax-like texture helps retain water.

Index

amphibians 16

animals 5, 8, 12, 16, 20

birds 8

canopy layer 11, 12

emergent layer 6, 8

forest floor 19, 20

fruit 11

insects 8, 20

invertebrates 20

mammals 8, 16, 12, 20

plants 5, 6, 11, 15, 19

reptiles 16

sunlight 6, 15

understory layer 15, 16

Online Resources

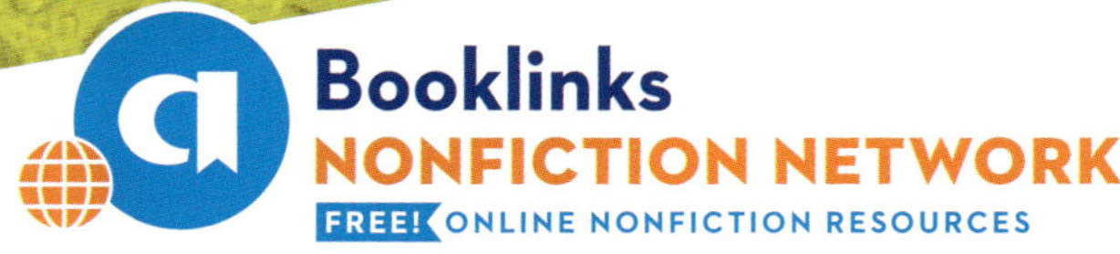

To learn more about layers of the rain forest, please visit **abdobooklinks.com** or scan this QR code. These links are routinely monitored and updated to provide the most current information available.